青藤手记

徐渭故里城市更新与改造实录

胡慧峰 主编

中国建筑工业出版社

图书在版编目（CIP）数据

青藤手记：徐渭故里城市更新与改造实录 / 胡慧峰主编. ---北京：中国建筑工业出版社，2024.8

ISBN 978-7-112-30212-3

I. TU984.255.3

中国国家版本馆CIP数据核字第20249C1Y11号

责任编辑：唐 旭
文字编辑：孙 硕
责任校对：王 烨

青藤手记 徐渭故里城市更新与改造实录
胡慧峰 主编

*

中国建筑工业出版社出版、发行（北京海淀三里河路9号）
各地新华书店、建筑书店经销
天津裕同印刷有限公司印刷

*

开本：787毫米×1092毫米 1/32 印张：5$^{1}/_{8}$ 字数：115千字
2024年9月第一版 2024年9月第一次印刷
定价：76.00元
ISBN 978-7-112-30212-3
(43562)

版权所有 翻印必究
如有内容及印装质量问题，请与本社读者服务中心联系
电话：（010）58337283 QQ：2885381756
(地址：北京海淀三里河路9号中国建筑工业出版社604室 邮政编码：100037)

主编单位
浙江大学建筑设计研究院 建筑创作研究中心

主　编
胡慧峰

执行主编
蒋兰兰

编　委
(根据文章先后顺序排列)
赫英爽　宋　羽　章晨帆　韩立帆　李鹏飞　陈赟强

摄　影
(根据文章先后顺序排列)
韩立帆　宋　羽　章晨帆　赫英爽　蒋兰兰
王小冬　李鹏飞　陈赟强　赵　强　雷坛坛

美术编辑
余舒烨

青藤手记
by UAD ACRC

前言 FOREWORD

"又见青藤"是针对青藤片区的"城市更新计划",是浙江大学建筑设计研究院建筑创作研究中心(UAD ACRC)在绍兴古城改造行动中的重要实践。在设计理论和建筑形态错综复杂的当下,语言和符号正在丧失意义,而权力与空间的博弈正逐渐加剧城市的模糊属性。"又见青藤"策划案,试图打破这些桎梏,跳出自上而下的物质空间改造和贫乏的官方文化叙事,挖掘历史街区中被遗忘的片段,从多重尺度介入古城的保护与更新,将历史变成资源,给予城市新的活力与创意。通过现代实践"干涉"线性时间秩序下的古城生长体系,将历史街区中突出的普遍价值所捍卫的整体性和本真性具象化表达,激发街区的空间想象力,让传统文化需求获得新的物质载体,赋予青藤片区更有生命力的定义;在平衡土地价值的同时,透过历史性城镇景观的视角,以人为本全方位提升青藤片区的街区幸福感。

"YOU JIAN QINGTENG", planned for urban renewal in Qingteng Community, is an indispensable strategy in the renovation of Shaoxing ancient city undertaken by UAD ACRC. At a time when design theories and architectural forms are intricate and dazzling, language and symbols are losing their meaning, and the game of power and space is gradually exacerbating a city's ambiguous nature. "YOU JIAN QINGTENG" attempts to break these shackles, to break away from the top-down space manufacturing and the tedious official narration of culture, to intervene in the protection and renewal of the ancient city from multiple scales by turning it into a Creative City with resources excavated from forgotten fragments of the historical district. The modern practice of "interfering" with the growth system of the ancient city under the linear time order expresses the integrity and authenticity defended by the Outstanding Universal Value of the historic district, inspiring the spatial imagination applied in the district, innovating material carriers for traditional cultural needs, and empowering Qingteng Community with a more vital definition. When balancing the value of the land, it enhances the sense of well-being of the neighbourhood in Qingteng Community through the perspective of the historic urban landscape in a human-centred way.

by UAD ACRC

目录 CONTENTS

前言
FOREWORD 4-5

第一章
CHAPTER 01
8-23

城市天际线
装备
晾晒肉干
硬核晒太阳
调研
张氏副食品店
发廊
立面更新

第四章
CHAPTER 04
56-73

城市面孔
锡箔纸
徐渭的照拂
集体记忆
烟酒茶糖
隔墙致意
绍兴味道
工人
防护窗森林

第七章
CHAPTER 07
108-125

街道起居
特立独行的墙
让食物更便捷
后窗
粉刷匠
夏日长
焕然如新
高楼记忆
盗梦青藤
雨中醉草

PROJECTS IN QINGTEGN SHAOXING
158-159

项目介绍

第二章 CHAPTER 02
24-39

见微知著
"将军"
指日可待
开工好天气
在现场
"茄子"
走亲戚
大隐隐于市

第三章 CHAPTER 03
40-55

盒饭
玻璃
夜空下
夕色春俏
角落里
友好
日常
幸福来敲门
门

第五章 CHAPTER 05
74-91

夜曲
貌合
哈哈
错过
借老
吃瓜
照面
路过
虚位
顾盼

第六章 CHAPTER 06
92-107

幻影
碰一个
狂而不乱
莲叶何田田
黑板魅力
夜游癖
坡道上的生活
始终很绍兴

第八章 CHAPTER 08
126-141

仓桥客厅
吉屋出租
旅途愉快
长颈鹿
榴花斋
优雅
你瞅啥

第九章 CHAPTER 09
142-157

看！"灰机"
美发
窗框再就业
地（上的）瓜
十秒熙攘
"其实我是猪"
恰如其分
日常问候

DESIGN GROUP
160-163

设计团队

EPILOGUE
164

后记

第一章
CHAPTER
01

城市天际线
装备
晾晒肉干
硬核晒太阳
调研
张氏副食品店
发廊
立面更新

SKYLINE PROHIBITION
OUTFIT
DRYING
HARDCORE SUNBATHING
SURVEY
ZHANG'S GROCERY STORE
BARBERSHOP
FACADE RENOVATION

青藤工地 | 2020-09-22

第一章 CHAPTER 01

2021年，胡慧峰老师带领 UAD ACRC 小伙伴开启"**青藤手记**"，记录和分享"**又见青藤**"城市保护与更新过程中的变化细节。

作为一个被90后"攻占"的团队，我们选择借助 PECHA KUCHA 的图文模式，用简单直白的方式呈现"**野生现场**"，并赋予它一个高级又时尚的职责——**用图像记录历史（图像口述史）**。

这里，有建筑师的驻场日常，有青藤居民的遛狗生活，也有项目的施工现场——它的琐碎和粗糙，让记录充斥凌乱感，而更显出城市烟火气的珍贵。

对！这就是我们对"**让设计落地**"最真实和接地气的表达。

2021, Hu is able to lead UAD ACRC team to get 'Notes on QingTeng Community' started, recording and sharing details of the vicissitudes of the city during 'YOU JIAN QINGTENG' project.

The team, conquered by the post-90s, chose Pecha Kucha presentation to display the life and stories around construction sites in a simple and straightforward way, but also vested it with a noble and fancy responsibility of visual recording as being the videotapes and transcriptions in oral history.

Audience could find scenes like architects' daily life, Qingteng residents' dog-walking and the construction condition. The triviality and roughness make the record full of clutter but activated in the preciousness of real urban life.

This is exactly the most down to earth way to turn our design into actual fruition.

城市天际线
SKYLINE PROHIBITION

城市天际线作为城市风貌的重要组成部分，其重要性和地位正在不断加强。欧美及日本等很多国家及地区都有对历史保护城市／区域的天际线发布过相关的规定和限制，以确保城市风貌的原真性和完整性。

但是，面临城市更新中诸如居住条件恶劣亟待改善等实际问题，天际线与城市高度是否应该被放置在更广泛的讨论议题中被重新考量，值得专业人士审慎思考。

青藤片区 | 2021-01-20

装备
OUTFIT

家门口晒太阳的装备展示。

其反映了社区居民的生活状态，也暗含诸如年龄分布等细节信息，是城市更新中不容忽视的重要内容。

青藤片区 | 2021-01-20

晾晒肉干
DRYING

年前,青藤片区居民开始晾晒肉干, 他们利用一切可以挂肉的工具。

肉——作为街道的"城市家具",被悬挂在晾衣架上、灯管的金属安装架上等,肉色鲜艳,构成独特的风景线。

仓桥直街 | 2021-01-19

晾晒酱肉对于北方城市长大的人而言，和挂在室外五颜六色的内衣裤一样，都是陌生而新鲜的事。所以，当平日里四处悬挂的鲜艳衣物被沉甸甸的酱肉取代时，这些重新占有街道空间的"新物种"，让城市瞬间变得更加真实而有温度。

过年的气息藏在居民对阳光的期盼中，因这些露脸的"肉们"而生动起来。我看到的，也终于不仅是作为"旅游城市"的绍兴，还是沾染了生活气息的街道片区。

晾晒酱肉的行为所赋予街道空间的独特性，大概需要被纳入青藤片区营造中，成为一个微小却也不可或缺的组成因素。

To us, the person growing up in North China, drying the cured pork is as surprisingly new as hanging colorful underwear outside. When the daily hung bright clothes are replaced by these heavy sauced new objects that reoccupy the street space, the city tends to be more real and warm in an instant.

Wrapped in residents' expectation of the sunlight lies the approaching the Spring Festival triggered by these 'flesh' who show their faces. Finally, the Shaoxing city I saw is not only a tourist city, but street areas of spiced mundane life.

The uniqueness of the street space for drying cured pork probably needs to be incorporated into the construction of the Qingteng Community as a small but indispensable component.

硬核晒太阳
HARDCORE SUNBATHING

徐渭艺术馆项目正在进行施工,住在后观巷附近的阿姨粉粉嫩嫩地坐在四处凌乱的施工现场晒太阳,与过路的邻居大叔闲聊。

大叔正望向艺术馆施工现场。

后观巷 | 2021-01-19

调研
SURVEY

对社区现状的实时跟进记录是深入了解青藤片区的重要方法之一。

同时作为一手资料,也为片区改造项目的顺利进行提供更有说服力的数据支持。

青藤片区 | 2021-01-21

张氏副食品店
ZHANG'S GROCERY STORE

曾经的张氏副食品店正在经历"改头换面"。

仓桥直街后观巷路口 | 2021-01-21

阿姨的小卖部原来位于后观巷1号,因后观巷立面提升迁到此处。店后是阿姨的住处和厨卫,本不大的进深却大部分被柴暗笼罩。这是一处极具商业价值的显赫店址,将来定是商家必争之地,就像国金中心之于陆家嘴,凯德广场之于新街口。面对镜头,阿姨总是微微咧着嘴,向我们娓娓谈及此处的历史,淡定中的三分得意似乎是在无意中拉踩着周围的一众小贩,直到对这称霸商区的美好愿望被我打破——两瓶农夫山泉。我接过水转身出门,回头定格了这一瞬间,大概下次来买水的时候只能喝依云了。

The grocery used to be located at No.1 Houguan Xiang and moved here due to the facade elevation there. Behind the shop is the old grocer's residence, kitchen and bathroom, most of which are shrouded in darkness. This is an advantaged location with great commercial value, which will be a must for merchants in the future, just like IFC is to Lujiazui in Shanghai and CapitaLand Plaza is to Xinjiekou in Nanjing. Facing the camera, the old grocer lady always grinned slightly and told us the history of this site. Taces of pride with her calmness seemed to be inadvertently scorning on the surrounding peddlers. Her dream of dominating the business district was interrupt by me, who asked for two bottles of Nongfu Spring. I took the water, turned to go out and photographed this moment. Maybe next time I came to buy water, I could only drink Evian.

发廊
BARBERSHOP

作为社区的重要组成部分,记录社区居民的生活现状也是我们的工作内容之一。

徐渭艺术馆工地 | 2021-01-21

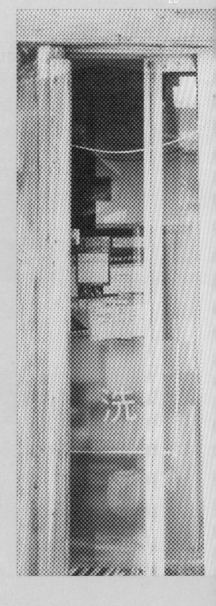

立面更新
FACADE RENOVATION

后观巷立面更新工程正在进行中。

街景是人们沿街道看向一排建筑物和设施的视角,其影响着商业的成功、车辆和行人的流量,以及社会互动的机会。

因此,精心设计的街景可以提升物业价值,提高安全性,让行人和骑行者感受到需求被重视,并能反映一个街区的独特性。

后观巷 | 2021-01-21

第二章
CHAPTER
02

见微知著
"将军"
指日可待
开工好天气
在现场
"茄子"
走亲戚
大隐隐于市

青藤工地 | 2021-02-13

第二章 CHAPTER 02

时值春节假期,徐渭艺术馆的施工也进入到最后的冲刺阶段。所有参与项目施工的工作人员可谓是在年味里带着节奏加紧工作。UAD ACRC 的驻场建筑师在现场服务的同时,用图像记录下了这段珍贵的回忆。

与此同时,在针对青藤片区居住情况进行的调研中,我们发现该区域内居住人口**普遍老龄化**。而仓桥直街上时而出现的年轻人携家带口看望父母公婆的场景,似乎也侧面印证了这一现象。可见,社区建设中的居家养老需求日益强烈,而**如何构建老年友好型社区**更是成为设计行业中的热门话题。

本期的手记帮助我们记录下与之相关的线索。

It was the Spring Festival holiday that the construction of Xu Wei Art Museum headed into the home stretch. All the staff of the construction project experienced a rush to the finish line during the holiday. ACRC's resident architects recorded this precious memory with images while serving on site.

Meanwhile, the investigation of the living conditions in Qingteng community shows that residents here are generally aging. The occasional scenes of young people visiting their parents and in-laws with their families on the Cangqiaozhi Street seem to confirm this phenomenon. The demand for home-based care for the elderly in community construction is coming on strong, and how to build an elderly-friendly community has become a hot topic in the design industry.

This time, the notes helps us record clues on it.

见微知著
MULTUM IN PARVO

正月初二,工人们已陆续回到工地,为艺术馆不久后的正式落成继续奋斗。

不得闲的还有驻场建筑师,他们需要及时跟进工程进度,积极配合施工方,保证工程质量。

而此时,艺术馆的轮廓已初见端倪。

徐渭艺术馆工地 | 2021-02-13

古城的午后,冬日的暖阳透过树枝,洒落在巷口的小桌板上。工地"咚咚咚"的轰鸣声丝毫不影响老百姓打"包红心"的闲暇生活。四周居民驻足观看,无意识间,这里就是一个社区活动场所。

他走了,他来了,他们还在。

建筑学领域中,场所的形成途径之一,便得益于一定数量的人群在同一地点形成定期的聚集和消散,逐渐成形的场所自带空间气质,进而带动形成具有鲜明特征的场域,而社区的灵魂往往也得益于此,通俗地讲,我们称之为"烟火味"。古城的保护、发展和利用,需要留存这份烟火味。

In the afternoon of the old town, the warm winter sun shines through the branches and falls on the small table at the entrance of the alley, and the noise of the construction site does not affect the leisure life of ordinary people playing 'Bao Hong Xin' (a special card game popular in Anji, Zhejiang Province.) at all. Soon after, it attracted many residents around to stop and watch, and here becomes a community activity place.

One man coming, the other man going, they're still here.

In the field of architecture, one of the ways to form a place benefits from the regular gathering and leaving of a certain number of people in the same place. The gradually formed place has its own feature, which leads to the formation of a field with distinctive characteristics, as well as the soul of the community. Basically, we call it 'the taste of earthly life'. The protection, development and utilization of the old town need to retain the taste.

"将军"
ENTERTAIN

没有场所，就制造场所。"野生的"公共空间在青藤片区改造期间几乎随处可见。事实上，类似的"棋局"在绍兴市乃至全国都屡见不鲜。作家双雪涛就曾在短篇小说《大师》中生动描写过一位生活在东北、棋技一流的父亲角色。同时可见，老龄化社会是当前我们正在面临的重大社会议题。

因此，城市更新中如何为老年人提供更适宜的公共空间，是设计师亟待思考和解决的问题。

老吾老以及人之老，历史保护中"历史"的含义应当更加丰富。

青藤片区 | 2021-02-13

指日可待
ALMOST DONE

艺术馆建成指日可待。

与以往的熙熙攘攘不同,此时的后观巷异常安静。

古树生新芽,我们相信,青藤的未来会越来越好。

后观巷 | 2021-02-13

在现场
HU IS CONTEMPLATING

对于保障建筑质量而言，建筑师的现场服务是至关重要的环节。

因此，UAD ACRC 要求建筑师定期驻场，随时主动发现问题。同时，他们也需掌握随时"地"画图的本领。

徐渭艺术馆工地现场 | 2021-02-24

如果你经常去驻场，你会发现建筑工人往往会对一些穿戴整齐的人很敏感，并且富有警惕性，暗中去观察他们的行为，判断他们的身份，是项目的业主？驻场的建筑师？或是他们的工地经理？看到我举起了照相机，他大概是卸掉了防御，露出了笑容，镜头感也是非常强的。

事实上，每次去我也拍了很多驻场建筑师的照片，但他们大多都是紧缩眉头或若有所思的状态，我想，等建筑完工的那一天他们也会露出这样的笑容吧。

If you often go to the site, you will find that construction workers tend to keep a wary eye on some well-dressed people, observing their behavior secretly and judging their identities. Is the owner of the project? Resident architect? Or their site manager? Seeing me raising the camera, he probably took off his defenses and smiled. Then, a good photo is made.

In fact, I also took a lot of photos of the resident architects, but most of them were frowning or thinking. I think they will show such a smile when the construction is completed.

"茄子"
'CHEESE'

工人们除了上工，拍起照也是相当一流。

与小红书上令人"傻傻分不清楚"的照片相比，他们理应被更多地记录。

徐渭艺术馆工地 | 2021-02-24

走亲戚
AS HAPPY AS PEPPA PIG

元宵佳节将至,年轻夫妇牵着幼童"走街串巷"。

此时,里弄成为城市表情的容器:

"火红的灯笼"挂着人们对新一年美好愿景的期盼;牵在手中的"佩奇和小牛",倒更像是时代的符号。

年轻人与老房子的关系,也许有更多可能性。

仓桥直街 | 2021-02-24

大隐隐于市
A DECENT PROFILE

建筑体量与功能对空间面积的需求，一直以来，都是"在历史街区中建造一座美术馆"需要面对和解决的重要问题之一。

青藤片区 | 2021-02-13

第一卷
CHAPTER
03

盒饭
玻璃
夜空下
夕色春俏
角落里
友好
日常
幸福末班车

LUNCH
GLASS
STARRY NIGHT
SPRING TWILIGHT
MAN ON THE CORNER
LUCKY
EXPOSURE
THE PURSUIT OF HAPPYNESS
DOOR

徐渭艺术馆内部 | 2021-03-11

第三章　CHAPTER 03

春日料峭，清明将至，阴雨绵绵，青藤难得见日光。

步入收尾工作的艺术馆和广场终于初见端倪，繁忙的工地此刻回归昔日的安静。连同成堆的建筑材料、钢筋脚手架和巨大的吊车一起消失的，还有大量工人的身影。

此时正值午休，他们零星几个散落各处，手端盒饭就地进餐。日落时分，广场上三三两两的居民路过，除了牵在手中的小朋友，不知谁家的小狗也悠哉悠哉闪身而过。大乘弄上的青藤书屋正在确认室内展陈和家具摆放，溜上徐渭艺术馆二层，刚好一窥青藤书屋的真容。

这一期手记中，老年人的生活轨迹依然随处可见。除此之外，我们还偶遇了党建活动，并在斗折蛇行的民居缝隙中感知社区土著们的日常起居。

It rains all the time on such chilly spring days, so the green vine seldom sees sunlight. The Qingming Festival is approaching.

As workers put on the finishing touches to the construction of the art museum and square, the site has now returned to its former quiet. Along with piles of building materials, steel scaffolding and huge cranes, there are no sign of throngs of workers. At the lunch break, a few of them were scattered all over the place having lunch.

At sunset, one or two residents with their children are walking past the square. Also, some puppies walk past leisurely. In the Qingteng Study on Dacheng Lane, the staff are confirming the indoor exhibition and furniture arrangement. I sneak in the second floor of Xu Wei Art Museum, just to get a glimpse of the Qingteng Study.

In this note, the life of the elderly can still be seen here and there, and we also came across party construction activities. In addition, we felt the daily life of the indigenous people in the community through the winding path to residential buildings.

盒饭
LUNCH

徐渭艺术馆的施工接近尾声,青藤广场也终于一点点呈现出设计图纸上的样子。

青藤广场 | 2021-03-04

玻璃
GLASS

建筑的建成效果高度依赖于施工中对建材纹理、特征的严格把控,这不仅要求建筑师熟悉各种材料,也需要他们高素质、高品质的现场服务。

青藤工地 | 2021-03-04

夜空下
STARRY NIGHT

夜空下的徐渭艺术馆。

也许,梵高(VAN GOGH)笔下圣雷米(SAINT REMY)的星空,与这里共享着苍穹的秘密。

徐渭在如此夜色中,会看到什么?

徐渭艺术馆主入口前 | 2021-03-04

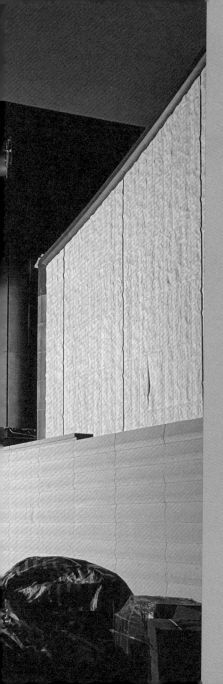

把灯调暗一点,鸟儿才能睡觉;把路修窄一点,人才能够相遇。

但你抵不住城市开发者的汹涌热情,和一场盛大而隆重的形式主义纪念。一切都曝光在华丽的灯光之下,众人都在喝彩,你却很费劲地在找月亮,多少有点荒诞。

设计师多少有点失心疯般患得患失,既见你高台稳筑、装饰华美,又求个浪漫诗意、小众调调,难得糊涂又难得清醒。

所以每一次都寄希望于下一次,和人生一样。

夜空下,微风吹来微风凉,吹来月亮和星空。

说声晚安,明天又是艳阳天。

Dim the lights so the birds can sleep. Narrow the road so people can meet.

But you can't resist the surging enthusiasm of city developers and a grand memorial of formalism. Everything is exposed under the gorgeous light, everyone is cheering, but you are struggling to find the moon. It's somewhat absurd.

Designers are somewhat a little bit frustrated and frantic about gains and losses. It's very hard to see high buildings decorated beautifully, and also with romantic poetic and niche tone.

So every time we hope for the next time, same as life.

Under the starry night sky, breeze was cool.

Good night! Tomorrow is another sunny day.

夕色春俏
SPRING TWILIGHT

夕阳西下,春色伶俐。
街区居民拉着小朋友站在未来的青藤广场上眺望徐渭艺术馆。

青藤广场 | 2021-03-04

角落里
MAN ON THE CORNER

艺术馆旁边的青藤书院,工人站在角落里举着手机。

青藤书院 | 2021-03-11

这是一个裸露的空间,没有任何材料对这个空间进行塑造,未来这个空间将是最开放的,细微的层高差异创造了一定的舞台效应,框住了这个空间的人,"看"与"被看"。

但现在的这里是最私密的,右侧的"施工场地,禁止入内"给予了建筑工人专属的空间,我站在外面好像看到了一个听不见的故事,但好像又能感受到他悄悄地在和家人通话。它激发了一种现实与幻影的相互作用,为这场景添加了一抹新的视觉维度。

This is an exposed space, which has not been shaped by any materials yet. In the future, this space will be the most open. The subtle difference in floor height creates a certain stage effect, which frames the people in this space, the role of 'seeing' and 'being seen'.

But now this place is the most private. The 'Construction site, no entry' on the right side allows the construction workers to have exclusive space. Standing outside, I seem to watch an inaudible story, but I can perceive that the worker is having a telephone conversation with his family. It inspires a kind of interaction between reality and phantom, adding a new visual dimension to the scene.

友好
LOCALITY

调研中,我们经常会看到老年人的身影。他们的生活痕迹从居住的屋子蔓延到街道、河边和广场。

吃饭、聊天、买菜、遛狗,抑或如图片中的爷爷,坐在家门口望着街道上的行人。「门」这一建筑元素,在片区中常常呈现打开/半打开状态。

面对镜头露出笑容而从无疑虑,他们对陌生人表达的友好,让我们看到「青藤老人」的典型面孔。

青藤片区 | 2021-03-11

日常
EXPOSURE

当日常被毫无保留地曝露于日光之下，我们看到生活的真实、凌乱和琐碎。它们与通常定义中的"美好整齐"相去甚远，但它们不重要吗？

这些基础设施，在城市更新中往往会被更加舒适好用的标准化产品代替。

之后，隐藏在图片中的"生活气息"会否随之消失？古城的独特性和完整性究竟如何，以及怎样附着在物理空间中？

青藤片区 | 2021-03-11

幸福来敲门
THE PURSUIT OF HAPPYNESS

作揖坊位于仓桥直街以北,两街平行,中间隔着河道。拍摄当日,此街一处台门内正在进行党建活动。

牌匾上的"台门党校"黑底红字,很有气势。

大门敞开,人群四散于各处,熙来攘往也是热闹。

小伙子举着摄像机认真工作,台门内的街坊邻居也都兴高采烈,瞧着院子里正在进行的采访活动。

龙山后街 | 2021-03-11

门
DOOR

后观巷以北的一处院子里,有这样一排态度鲜明的房子。立面上一共五扇门,无窗。门之间彼此保持着极其随意的间距,同时共享着"大门紧锁"的统一表情。

青藤片区(后观巷以北) |2021-03-11

第四章
CHAPTER
04

城市面孔
锡箔纸
徐渭的照拂
集体记忆
烟酒茶糖
隔墙致意
绍兴味道
工人
防护窗森林

URBAN COMPLEXITY
MEMORIAL
HIGH FIVE IN THE AIR
COLLECTIVE MEMORY
ALCOHOL & TOBACCO
REBIRTH IN RUIN
A BITE OF SHAOXING
WORKERS
FENCED WINDOW

居民家门上的通知 | 2021-04-07

第四章 CHAPTER 04

社区 (COMMUNITY),最早由19世纪德国社会学家滕尼斯提出,指与"社会"(GESELLSCHAFT) 相对的通过血缘、邻里和朋友关系建立起来的人群组合;后亦指:因为共享共同价值观或文化的人群,居住于同一区域,以及从而衍生的互动影响,而聚集在一起的社会单位。

建筑领域中针对"社区"的讨论通常与"公共空间"关联,因为它是表达"共享价值观"的最优解。**而此次,我们尝试切换角度,通过对集体记忆 (COLLECTIVE MEMORY) 的图像记录,讨论在"平衡古城保护与更新"语境下的社区未来面孔。**

正如法国历史学家皮埃尔·诺哈(PIERRE NORA)在研究地方与空间中提到的:"一个'记忆的场所'是任何重要的东西,不论它是物质或非物质的,由于人们的意愿或者时代的洗礼而变成一个群体的记忆遗产中标志性的元素。"青藤片区中的老墙、民居形态、锡箔纸、烟酒牌,以及新建的徐渭像等,共同构成对城市记忆的描摹,为城市历史的延续提供有力证词。

The concept of Community (Gemeinschaft), first proposed by the German sociologist Ferdinand Tonnies in 1920s, corresponded with 'Society' (Gesellschaft), refers to the group of people whose closeness of relationships are blood, neighbors, and friends. Later, it is used as a social unit to groups whose interactive influence are from sharing common values or culture when living in the same area.

When it comes to 'community' in discussions in the field of architecture, it is usually associated with 'public space' as the optimal solution for expressing 'shared values'. This time, with attempt to change perspectives, we discuss the future appearance of the community in the context of 'balancing protection and renewal of ancient cities' through image recording of collective memory.

As the French historian Pierre Nora mentioned in the study of place and space, namely Lieux de memoire-Realms of memory, that the site of memory is any significant entity, whether material or non-material in nature, which by dint of human will or the work of time, has become a symbolic element of the memorial heritage of any community. In Qingteng community, old walls, residential lives, tin foil paper, entertainments like tobacco liquor and cards, and the newly-built Xu Wei statue together constitute a memory description of the city and testimonies for the continuation of the city's history.

城市面孔
URBAN COMPLEXITY

多元城市的体现之一,即在于其建筑形式的丰富姿态——不同年代的房屋共同构成统一而完整的古城意向。

它绝非呆板地复制粘贴 A,也避免刻意地突兀造作。

肌理的生成反映了城市设计师对"设计语言"的深入思考与精准应用。

青羊片区鸟瞰 | 2021-04-08

锡箔纸
MEMORIAL

正值清明,仓桥直街上的小铺子门口纷纷挂起扫墓用的锡箔纸。

与腊肉和晾晒衣物相似的陈列方式"挂"体现出一种态度和生活方式,成为街道上的"临时展品",在传统节日的轮回中记录时间的流逝。

仓桥直街南段 | 2021-04-07

我对墙有种奇怪的执念，因为能够想象的场景很多，晚风中匆匆路过的人影，低头靠着抽烟的青年，以及随意坐在墙根下打盹的大爷。

这堵墙更加神秘，不知道是哪一年，原本可以随手推开的窗户外，又砌起了另一堵高墙。关起来的是浪漫旖旎的心思或是对新生活蠢蠢欲动的幻想。

2021 年，窗内窗外的人事又变更了一个轮回，路过的姑娘也许早嫁了人，窗内的工人下岗又退休过上了安逸的晚年生活，停下拍照的也不知从何处慕名而来。挺好的，刮风下雨你也没倒。还能再看 100 年的笑话。

I have a strange obsession with the walls, for it triggers me to image countless scenes, such as the silhouettes of people passing by in the evening breeze, the young people smoking against the wall with droopy heads, and the ole man sittings casually and dozing under the wall. This wall is even more mysterious when one year another high wall was built outside the window that could have opened any time with little push. What has been locked up is romantic and charming thoughts or fantasies about a new life.

In 2021, the vicissitudes inside and outside the window has drifted away. The girl passing by may have already married someone, and the worker inside the window may have had a comfortable retirement after being laid off. Those who stopped to take pictures were new to this 'famous' city. It stands still after years of winds and rains and can still watch well-beings of this world more than 100 years.

烟酒茶糖
ALCOHOL & TOBACCO

林语堂在《人生的归宿》中说:"在中国,精神的价值非但从未与物质的价值相分离,反而帮助人们尽情地享受自己命里注定的生活。"

在青藤,杂货铺卖烟酒,小馆子卖烟酒,发廊卖烟酒,按摩房更是卖着烟酒。

后观巷 | 2021-02-24

你好,陌生人。

你斜靠在旧时的月色里,我伫立于高悬的氙气大灯下。很难说,是我的镜头入侵了你带着几分落寞的闲散,还是你闯进了我对这城市进程的偶然观察。

墙这边,岁月无言淘洗日子。
墙那头,城市轰然生长。
这一刻,城市记忆脱离集体的轨道,
以个体的语言,
说话。

Hey strangers,

You lean against the moonlight of the old time, and I stand under the high xenon headlights. It's hard to say it is whether my camera lens invade your somewhat lonely idleness, or you break into my accidental observation of the city development.

On this side of the wall, days sluiced out of years in silence.
On the other side of the wall, the city grows in sudden.
At this moment, the memory of the city is out of the collective track,
In individual language,
it speaks.

隔墙致意
REBIRTH IN RUIN

施工中的绍兴师爷馆,与一旁的民居隔墙相望。

城市记忆此刻正停留在这堵拆迁中的墙体之上,勾勒出青藤街区「新旧」对峙与互动的暧昧特征。

绍兴师爷馆工地 | 2021-03-31

绍兴味道
A BITE OF SHAOXING

一位路过的大叔特意叫住举着照相机的我,举起酒瓶,摆好姿势。

"来!来!拍这个!但别把我拍进去。"他特意强调。

仓桥直街南段 | 2021-04-07

工人
WORKERS

阳光下的工人。

徐渭艺术馆 | 2021-03-13

防护窗森林
FENCED WINDOW

在大面积清末民初的单层民居中,也不乏图中的居民楼建筑。

该类型民居兴盛于20世纪90年代初期,白色瓷砖饰面,6~7层,无电梯。

并且,家家户户安装防护窗。

青藤片区 | 2021-04-07

第五章

CHAPTER

05

夜曲
貌合
哈哈
错过
偕老
吃瓜
照面
路过
虚位
顾盼

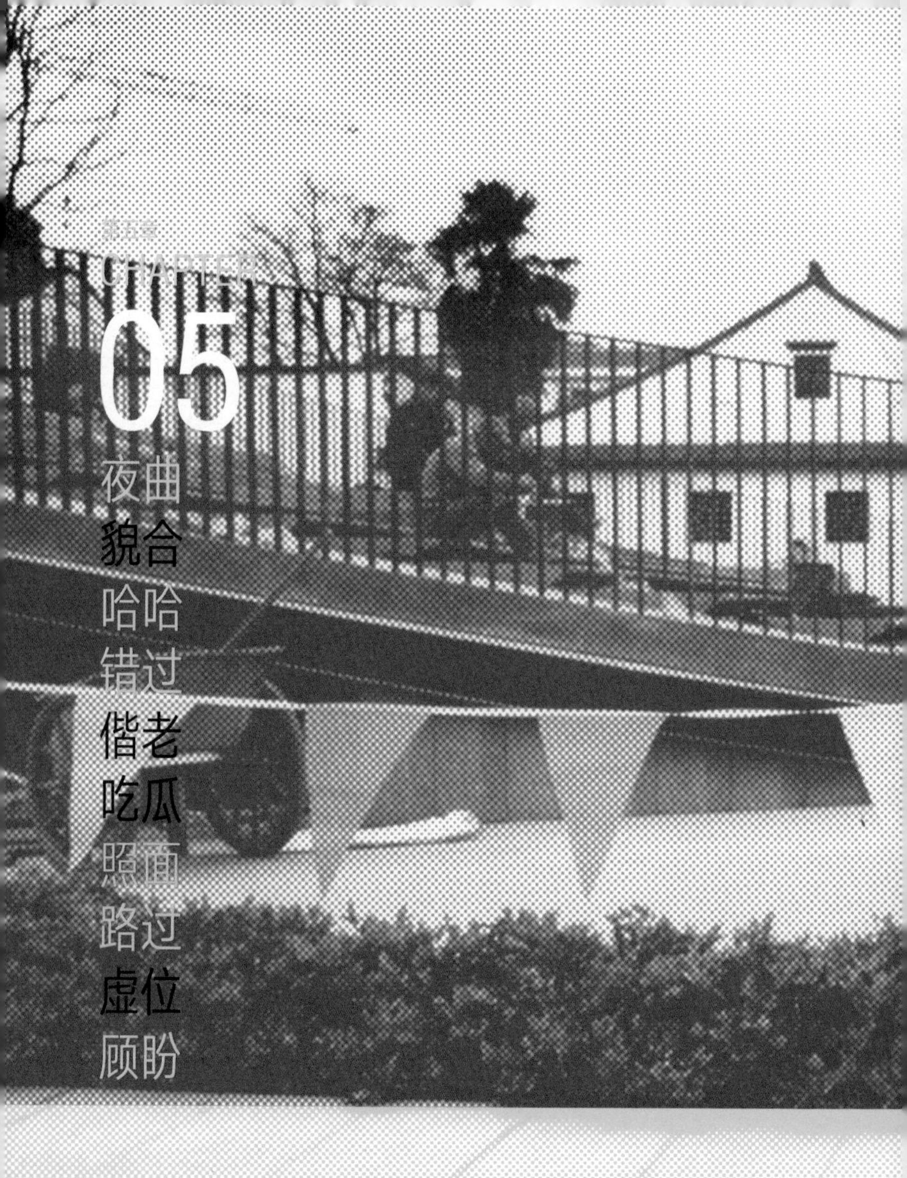

HALF MOON SERENADE
VARIANCE
HAHA
MISS
FOREVER
SPILL THE TEA
ENCOUNTER
PASSING BY
VACANT FOR
LOOK FORWARD TO

青藤片区马路边|2021-05-18

第五章 CHAPTER 05

终于，徐渭故里正式对公众开放，徐渭艺术馆也要隆重和大家见面了。刚好，在"520"前一天，老胡带着 UAD ACRC 的伙伴们一道去青藤片区参加典礼！

虽说 520 只是个数字，但生活毕竟需要仪式感，建筑师们的画图日常在这个特殊的日子里，因奇人徐渭洒脱不羁的一生，也跟着一起浪漫起来。

其实所谓艺术社区，艺术与社区的关系理应被赋予更加丰富的可能性，它们之间也像一对从陌生到热恋的伴侣，初识的青涩单纯，别扭争吵，到逐渐熟络，相知相伴。艺术投身社区的天马行空和肆意妄为，在家长里短的闲篇琐碎中触摸最真实的人、真实的生活和真实的世界。如此，青藤才能常青。

At last, the former residence of Xu Wei is officially open to the public and the Xu Wei Art Museum is going to say hi to everyone. On the day before 20 May, as the UAD ACRC team is lead to the ceremony, we get the chance of rubbernecking!

Semiotically, 520 is just a figure, but it might light the life with a sense of ritual (520 and 'I love you' are homonym in Chinese). The drawing routine of single architects seems to be somehow romantic on this special day, after knowing the uninhibited life of Xu Wei.

In fact, the relationship between art and the community should see richer possibilities, for they are like a pair of partners who have gone from strangers to lovers, going through the youthful innocence of the first acquaintance, the twisting and quarrelling, to their gradual acquaintance and companionship. The art to the community ought to be wild, touching the real people, the real life and the real world in the casual and trivial stories of families. In this way, the vine of Qingteng Community will always be green.

夜曲
HALF MOON · SERENADE

小姐姐们的曲子真好听。

所谓"琴棋书画诗酒花,柴米油盐酱醋茶。"

艺术和社区的关系都在曲子里,也都在诗里。

徐渭艺术馆二层中庭 | 2021-05-19

貌合
VARIANCE

"这玻璃水……""光线很好……""……的做工感觉还行。""哎,对面有个漂亮小姐姐!""都好都好。""我手机里那张效果图呢?怎么找不到了……""你瞅哪儿呢?""这也能上微博头条,也太敷衍了。"

徐渭艺术馆内二层中庭 | 2021-05-19

哈哈
HAHA

成年人的世界,开心也好,思虑也罢,好像都能哈哈而过。

徐渭艺术馆展厅 | 2021-05-19

错过
MISS

兰:"只要我跑得足够快,我就是艺术馆里的一道黑影。"
羽:"行吧,但你知道么,刚刚你已经错过了好几位大叔……"

徐渭艺术馆内二层中庭 | 2021-05-19

总觉得理想中感情的归属，可能就是这种归于平常的陪伴。年轻时总觉得爱情得轰轰烈烈，少说得有些离愁别恨，才能品出点一二三四，所以总不理解父母的柴米油盐，磕磕碰碰。等到自己结了婚，仿佛四目清朗，醍醐灌顶，既懊恼年少无知，又隐约心生期待，十足矛盾。但日子还是如水过去，在争吵中过去，在开心中也过去，留给每个人的时间都一样多又一样少。可能到了最后，也不过渴求一份手牵手的默契。在美术馆、在山野间、在寻常街道、在世界的哪里都好，回头有人慢慢跟着，吐槽有人嘿嘿笑着，好像确实也挺不错。

I always think the idealbelonging of affection may be such ordinary companionship. When I was young, myperception was always that love was so vigorous that parting and grievanceswould be must to taste the feeling of love. That is why I could not understandparents' routines. When I got married, I seemed to be enlightened when reachingan epiphany, full ofambivalence-not only annoyed by my youth and ignorance, but also vaguely expectingsomething. As life can be a glimpse of eyes, whether in the quarrel or in thejoy, my wife's time is as equal as mine. Maybe when approaching the end, what Idesire is just an intuitively convention of holding each other's hands. Itcould happen in Xu Wei Art Museums, in the mountains, in ordinary streets, andany other places in the world. What matters is to see being slowly followedwhen turning around and to be echoed with laughter when joking and teasing,which is pretty nice.

对附近居民而言,艺术馆如约开放当天,注定是吃瓜的好日子。他们围在南广场靠近青藤书屋的一排民居下,隔着临时搭起的围挡兴致勃勃地"看戏"。当然,这也是一睹附近常住民真容的好时机。整体而言,老年人居多。

在社区营造中,"如何提高居民参与度"是包括政府工作人员、设计师和非政府组织(NGO)都非常关注的话题。我们急需摆脱粗暴的一概而论,摆脱将"老年人参与度高"简单归结为"他们清闲、不上班,无事可做";相反,我们应挖掘背后更深层的原因,比如"为何老年人无事可做?社会是否提供他们可以融入社会生活的其他方式?为何年轻人参与社区活动更少?相关的社区活动内容是否能够吸引他们迈出家门?以及居民对于社区以及社区活动的需求究竟是什么等问题。

美术馆的开放无疑会给青藤带来巨大改变,希望这些变化能够让青藤焕发出更持久的活力。

The day the art museum opened asscheduled was destined to be a good day for nearby residents to be rubbernecks.They surrounded a row of residential houses in the South Square near theQingteng Study Room, and were enthusiastically "watching the show." over thetemporary fence. Vice versa, this was also a good time to have a glimpse of thoseresidents. In general, the majority were the elders.

Incommunity empowerment, "how to improve residents'participation" is a topic ofgreat concern to government officials, designers and NGOs. We are in urgentneed to get rid of harsh generalizations, as well as simply ascribing "theelders are highly engaged" to "they are idle, not working, and have nothing todo". Instead, we should decipher reasons furthermore by considering questionslike "Why the elders have nothing to do? What can be done? Does society provideother ways for them to integrate into social life?" "Why are young people lessinvolved in community activities? Can relevant community activities attract youngstersout of their homes?" and "What are residents' needs for community and communityactivities?"

The opening of the museum willundoubtedly change Qingteng Community, and may these changes make Qingteng Community glow with ever-lasting vitality.

照面
ENCOUNTER

开幕典礼前一天的青藤广场上,和墨镜大爷打了个照面。

大叔手机插胸口,单车上接送小孩的座椅和买菜筐子样样皆备。他很酷地盯了我一秒,很酷地扫视了广场一周,然后转身走了。

徐渭艺术馆南门入口 | 2021-05-18

虚位
VACATION FOR

最有趣的是大战前夕的平静,连风都透着紧张,椅子都毕恭毕敬,严肃到好笑。

我端着相机,听着软糯不明其意的绍兴话,很是紧张,广角都不太会用了。

青藤广场 | 2021-05-19

顾盼
LOOK FORWARD TO

左顾右盼,七上八下。静坐遐想,青藤幽深的巷弄走过几名妙龄姑娘。

慌慌张张,手足无措。午后雨中,街角的咖啡馆子,坐着几个聊赖青年。

青藤片区 OOOH 咖啡店 | 2021-05-19

CHAPTER 06

幻影
碰一个
狂而不乱
莲叶何田田
黑板魅力
夜游癖
坡顶上的非洲
始终很绍兴

PHANTOM
CHEERS
XU WEI'S CALLIGRAPHY
LOTUS LEAVES
HANDWRITING
NOCTURNALISM
SILHOUETTE OF YOU
BE SHAOXING ALWAYS

徐渭艺术馆门口|2021-07-06

第六章 CHAPTER 06

情人节特辑发布至今,已近两个月的时光。这期间的青藤片区经历着十足的大变化,艺术馆的正式营业为社区招揽来丰富的客群:从徐文长先生的资深画迷,到网红地打卡的忠实爱好者,以及成群结队、相伴而来的祖国未来的花骨朵儿们,相当老少皆宜。整体而言,到访者的年龄分布更趋于年轻化,曾经相对安静的传统社区也逐渐开始进入角色,成为名副其实的"徐渭故里"。

只可惜 UAD ACRC 人力有限,主力大多坐镇东二沉溺投标,其余则奔赴古城东南的大禹陵景区,为大禹品鉴会的顺利进行加班加点。因此,我们只能得空儿记录一些片段,或者在小红书等网络平台暗暗关注。比如我们发现仓桥直街上新开了咖啡店,老板藏身旁边的东鑫外贸。

总之,盛夏时节的青藤异常动人,做好防晒快去感受一下吧!

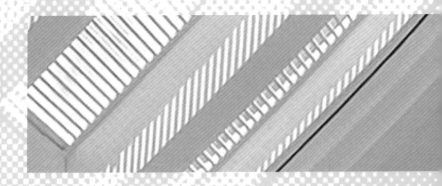

It has been nearly two months since the release of the Valentine's Day Special. During this period, the Qingteng Community has undergone great changes. The opening of the art gallery has brought in a rich audience from painting fanatics of Xu Wei, loyalists of social media posts, to children with promising future in groups. In general, the visitors tend to be younger and the community used to be relatively quiet and traditional is rightly recasting itself as 'former residence of Xu Wei'.

Unfortunately, the small UAD ACRC family have to either focus of the latest bidding, or go to the Dayu Mausoleum in the southeast of Shaoxing city to work overtime for the appreciation event. As a result, what happen in Qingteng Community could only be recorded in snippets by us or followed secretly on the social media RED. For example, we noticed that a new café opened on Cangqiaozhi Street and the owner was hiding in neighbour store selling foreign products.

All in all, the beauty of Qingteng Community is blossoming in the height of summer. It is worthy to feel the vitality with the protection from sun burn!

很少夜间逛青藤，看到这个场景，恍恍惚惚，空间重叠，仿佛下一秒，文长就要长衫翩翩、醉意朦胧地步入园中，又似乎儿时的夏日夜晚，坐在巷子口，依靠着姥姥，等着冰镇西瓜解暑。总觉得老底子的房子好，低低的檐口，顺着滴落的水滴就能看一下午，老底子的院子也好，老得不行的大树，伴着知了的叫声，能酣睡半晌。暖暖的灯光透过窗棂，朦胧的影子，是忙碌的母亲，偶尔听见低低的絮叨，太过温暖，让人莫名哽咽。许是自己也长大成人，庸庸碌碌，都不曾短暂停下来，感受生活了。

晚风中吹过，几帧从前闪过，在这道貌岸然的俗世里偷点儿时光，真真难得。

I rarely visit Qingteng at night.When I see this scene, I am in a trance. It seems that I am XU Wei, the greatscholar and artist, in a flowing long gown, slightly drunk, being about to stepinto the garden. It also seems that it is on a summer night when I was a child,sitting at the alleyway, relying on my grandma, waiting for the iced watermelonto cool the heat. I always think the old house is good. I can watch waterdroplets dripping along the low cornice for the whole afternoon. The old courtyardis also good. I can sleep soundly accompanied by the old tree where cicadas sing.Filtering the warm light, the window projects the hazy shadow of my busy motherwith occasional low ramble. It is too warm to make me stop chokinginexplicably. It might be due to the growing up as a mediocre. I never pausefor a while to experience life.

The evening wind blowing, a fewframes flashing past, it is really rare to steal some time from life in this specioussecular world.

幻影
PHANTOM

夜色中书屋里的光，
勾勒出院内墙面上斑驳的幻影
枯井天池、古树青藤

前观巷大乘弄 10 号青藤书屋 | 2021-07-04

碰一个
CHEERS

建筑师们大概很早就预想过广场建成后的一些使用场景，在特定的空间与时间中发出叮叮当当的声响。

青藤广场 | 2021-07-06

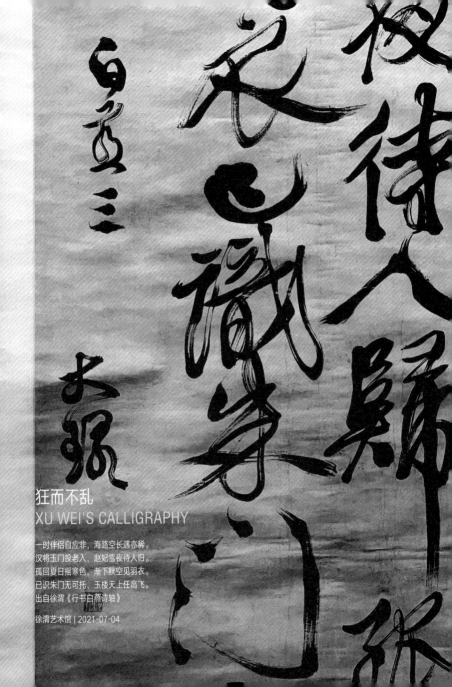

狂而不乱
XU WEI'S CALLIGRAPHY

一时伴侣自应非，海路空长遇亦稀。
汉将玉门投老入，赵妃雪夜待人归。
孤回夏日摇寒色，渐下秋空见羽衣。
已识朱门无可托，玉楼天上任高飞。
出自徐渭《行书白燕诗轴》

徐渭艺术馆 | 2021-07-04

莲叶何田田
LOTUS LEAVES

山光忽西落,池月渐东上。
散发乘夕凉,开轩卧闲敞。
荷风送香气,竹露滴清响。
欲取鸣琴弹,恨无知音赏。
感此怀故人,中宵劳梦想。
出自唐代孟浩然的《夏日南亭怀辛大》

青藤书屋老墙 | 2021-07-04

一块小黑板，看似渺小，却足以承载起一代人的生活记忆和情感交互，成为人与人交流的纽带；原住民看似为一个个独立的个体，聚合在一起成为社区，延续城市的生命。

一块小黑板，见证了社区的兴衰，见证了时代的变迁，如今却被遗忘在了角落，正如我们在尝试城市更新时常常遗忘在角落里却为社区注入活力的原住民。城市更新中需要在地性，需要关注社区里的原住民。

希望若干年后，它依然能在这里，记录生活的点点滴滴。

A small blackboard seems small, but is enough to embody life memories and emotional interactions of a generation, and becomes a communication link between people; the locals seem to be independent individuals, converging to a community, extending the life of the city.

A small blackboard, witnessing the rise and fall of the community and the changes of the times, is now forgotten in the corner, just like the locals who are often forgotten in the corner during urban renewal but plant vitality in the community. Locality is required in urban renewal, and attention needs to be paid to the indigenous locals in the community.

I hope that in a few years, it will still be here and record every bit of life.

黑板魅力
HANDWRITING

2021年五一，艺术家邱志杰在北京三源里菜市场做的"民以食为天"的艺术计划，让我们看到了书写的魅力。

黑板上的字：写了擦，擦了写。看字，也看写字的人，生活的细节都在字里，也是蛮不错的。

仓桥直街与和平弄交界 | 2021-07-02

以从脚下起步

休以你我似走

夜游癖
NOCTURNALISM

少年时恐惧黑暗，觉得其中暗藏浑浊秘境。

如今经过书屋，只看到光影，

又或徐渭的灵魂偶尔也路过，打个照面问声好。

前观巷大乘弄 10 号青藤书屋 | 2021-07-04

第七章
CHAPTER
07

特立独行的照片
让食物更便捷
时尚
粉刷匠
夏日长
焕然如新
雨后天晴
盆梦青藤
雨中醉草

LI-LONG PARLOUR
UNIQUE WALL
TAKEAWAY CITY
PHUBBING
LITTLE PAINTER
ECHOES OF THE SUMMER
BLANK-LEAVING
HIGHRISE IMPRESSION
INCEPTION
SINGING IN THE RAIN

徐渭艺术馆中庭 | 2021-08-11

第七章 CHAPTER 07

立秋时分,盛夏的热气还未消散。身处阴郁天色之中的青藤片区,反倒很是安然闲适。

日益丰富的人流构成,在这片安静的区域中各得其所。仓桥直街上新开的咖啡店里,咖啡师正与两位客人热情攀谈,门口的快递员在三轮电动车子上悠悠然耍手机,还有翘着二郎腿陷在藤椅里耍手机的大爷,藏身玻璃柜后小憩的阿姨,行色匆忙一闪而过的外卖员,不紧不慢神色悠然的垃圾运输工和漂漂亮亮来看展的游客们。

师爷馆和榴花斋的施工依然有条不紊地进行着,身处放眼望去郁郁葱葱的绿意之中,城市和她的市民好像都在过着暑假呢。

At the beginning of autumn, the heat of summer has not yet subsided. In the midst of the gloomy skies, Qingteng Community is, nevertheless acclimatized, very peaceful and relaxed.

Such tranquility guides the increasingly diverse 'employees' of the area to find their places. In the new café on Cangqiaozhi Street, the barista is chatting warmly with two customers. The courier at the door is lying on his three-wheel motorcycle. Apart from these people, there is an elder gent swiping his phone in a rattan chair with his legs crossed, an aunt taking a nap behind a glass case, a delivery man in a hurry, a rubbish hauler working in relaxation and visitors to the exhibition.

The construction of the Pomegranate Food Studio and the Shiye Pavilion is still in orderly progress, and the city and its citizens seem to be having a summer holiday in the midst of the lush greenery.

111

街道起居
LI LONG PARLOUR

大雨将至未至,大爷悠然闲坐认真耍手机。

开元弄四下无人经过,
如此,小巷子成了他的私人领域,好生自在。

开元弄 | 2021-08-11

特立独行的墙
UNIQUE WALL

机床厂遗留的老墙面,与市政翻新后的民居立面,
两位"老朋友"在不远处玻璃幕墙的见证下,
相互致意。

徐渭艺术馆老墙 | 2021-05-20

摩天大楼越来越高、居住空间越来越密、生活期阈值越来越大、消费方式越来越快，疫情虽然让世界短暂地暂停，但高科技与大数据却越来越洞悉人性，甚至带有侵略性袭来。

原本计划完工的师爷馆框架依然裸露着，安静的工地因为一辆外卖车驶过多了一分生机，这里是青藤片区的交通要塞，城市的快速运转也从这个口子渗透到了青藤片区。悄然间，青藤的人群结构与消费方式都产生了很大的变化，可以感受到青藤是一个有创造力的地方，热情的、简单的、自发的、由上而下、由下而上的。连接青藤的纽带也越来越多样，是网红在徐渭艺术馆的一组照片、是大乘弄街角的咖啡店、是可以跳广场舞晒太阳的青藤广场、是给街区原住民送外卖和快速的电动车。城市很繁忙、科技很发达，但我们都只是希望生活越来越好。

Skyscrapers getting taller, living spaces getting denser, the threshold of life getting increasingly bigger and consumption patterns getting growing faster, the epidemic hastemporarily suspended the world, but high-tech and big data progressively insighthuman nature, even invade with aggression.

The frame of the museum ofprivate advisers, which was originally planned to be completed, is stillexposed. A takeaway car enlivens the quiet construction site. This is thetraffic fortress of Qingteng Community, and is also the gate where the rapidoperation of the city penetrated into, Quietly, Qingteng's population structureand consumption patterns have undergone great changes. You can feel that,topdown and bottom-up, Qingteng breeds creativity, enthusiasm, simpleness,spontaneity. Accordingly, connections towards Qingteng become more diverse. Itcould be a group of photos from online celebrities in Xu Wei Art Museum. Itcould be the coffee shop on the corner of Dacheng Lane. It could be theQingteng Square where people can dance and bask in the sun. It could bescooters of food delivery and express to locals in the neighborhood. Cities arebusy when technology is advanced. We all just want life to get better and better.

让食物更便捷
TAKEAWAY CITY

在这个时代，外卖员始终陪伴着城市，一道走进和经历这个充满未知的每一天。生活依然进行，太阳照常升起。

绍兴师爷馆 | 2021-08-13

后窗
PHUBBING

艺术馆里的两面通高大窗,模糊了空间的内与外。

每一次从窗里眺望城市,都能隐约感知到一份藏于其间、委婉却紧密的联结,它关乎一方水土与一方人。

徐渭艺术馆 | 2021-08-13

夏日长
ECHOES OF THE SUMMER

纵使每一次途经仓桥小店,我们都充满好奇。
奶奶偶尔在小憩,
偶尔与客人或者邻居聊天,
偶尔和孙女打牌。
她的一天总是那么清清爽爽,悠悠闲闲。

仓桥直街 | 2021-07-06

焕然如新
BLANK-LEAVING

每当我们看到古城墙面在城市翻新工程中被统一粉刷时,都会有人感叹古城味道因此而消失。

历史保护领域中的"老态价值"(ALTESWERT)指的就是这些"老化、老龄"的城市构成。

这些甚至不具有"岁月价值"(AGE VALUE)的"老东西"是否一定指涉城市发展中的消极因素,美国历史学家大卫·洛文塔尔(DAVID LOWENTHAL)就曾对此写过文章进行探讨和论述。

仓桥直街 | 2021-07-06

青藤社区公告栏

最初按下快门的一瞬间是想记录与青藤相接的城市关系,站在仓桥直街沿着后观巷从西往东望去,可以看到与青藤风貌完全不同的建设银行大楼。不同的空间维度,也产生了一街之隔的差异性空间,在熵值的叠加作用下,时间创造着新的空间,旧的记忆与新的产物相互作用下,让青藤悄然从青藤书屋的前院重新蔓延生长,甚至有一点燎爆。青藤相对于整个绍兴足足变大了一个尺度,但相较整个城市而言,它还是保守的,但慢一点,蛮好。

The moment I pressed the shutterat first, I wanted to record the relation between the city and Qingteng. Standing on Cangqiaozhi Street and looking from the west to the east alongHouguan Lane, I could see the China Construction Bank, whose appearance iscompletely different from Qingteng. Different spatial dimensions also produce differentiatedspaces across the street. Under the superposition of entropy, time creates newspace. Under the interaction of old memories and new products, the Qingtengquietly rebirth from the front yard of Qingteng Study Room and is even glamorized.Qingteng enlarges Shaoxing City. But compared with the whole city, Qingteng is still conservative. However, it is good to be slower.

盗梦青藤
INCEPTION

每个盗梦者都要有一个自己的图腾，

用来分辨现实与梦境，青藤的图腾是什么呢？

徐渭艺术馆二层 | 2021-08-11

雨中醉草
SINGING IN THE RAIN

今日与君饮一斗,卧龙山下人屠狗。
雨歇苍鹰唤晚晴,浅草黄芽寒兔走。
酒深耳热白日斜,笔满心雄不停手。
出自徐渭《与言君饮酒》

青藤广场 | 2021-04-20

第八章
CHAPTER
08

仓桥客厅
吉屋出租
旅途愉快
长颈鹿
榴花斋
优雅
你瞅啥

CANGQIAO PARLOUR
FOR RENT
ENJOY YOUR TRIP
GIRAFFE
LUFTHANZHAI
ELEGANCE
WHAT'S UR PROBLEM?

青藤广场 | 2021-08-11

第八章 CHAPTER 08

一些熟悉的名字消失了,另一些却回来了。当哈利·波特再一次让魔法觉醒时,诺贝尔却把文学奖颁给"有点陌生的"坦桑尼亚作家。好在姗姗来迟的桂花香并没有辜负我们对季节的期待,立秋也终于用连绵不断的雨水和突降的气温把残留的夏天赶走了。

也许生活就是这样,反复在陌生与熟悉之间横跳。只是,青藤跳跃的节奏总是很悠然:榴花斋和师爷馆基本竣工,徐渭艺术馆的玻璃也差不多换好。伴随着每日的闲聊观望、路过停留与喝茶看报,变化依然在悄咪咪地进行。

Some familiar names have disappeared, others have returned. When Harry Potter once again awakens the magic of literature, the Nobel Prize is awarded to a 'somewhat unfamiliar' author from Tanzania. Fortunately, the late arrival of the scent of Osmanthus did not disappoint our expectations of the season, and autumn finally chased away the annoying summer with continuous rain and sudden drops in temperature.

Perhaps it is the life, tuning between the unfamiliar and the familiar. The rhythm of Qingteng Community, however, is always a leisurely one. The projects of Pomegranate Food Studio and the Shiye Pavilion are almost complete, and the glass replacement in the Xu Wei Art Gallery has come to the end. Despite being slight to notice at once, those changes still take place quietly, along with the daily chit-chat, stopping by, tea drinking and newspapers reading.

其实客厅这个词可能并不恰当，毕竟如果回到历史中去，传统中国家庭对空间的使用并不会单独分出来一块作为客厅。所以仓桥直街上出现的场景，就是这里人们的生活日常。因为室内无法满足一边做点儿事，一边聊天，所以他们走出来使用街道。这个场景对室外的需求带着纯天然气质，与大城市里的商铺进驻背后的逻辑全不相同。因此，我无法想象如果单独做个所谓的活动室或者公共客厅就能满足他们目前的需求。而新型空间的介入，也许会彻底打破这里的原真性。不过换个角度想的话，原真性一定很重要吗？背后的叙事逻辑是否又携带了其他意图？

In fact, the word parlor may notbe appropriate. After all, if we go back to history, traditional Chinese houses never separate out space as a parlor. Therefore, the scene on Cangqiaozhi Street is the daily life of the people here. They go out and use the street because they couldn't do something and chat indoors. The demand for the outdoors in this scene is purely naturalistic, which is completely different from the logic of shop running in big cities. Therefore, I can't imagine if asocalled activity room or communal parlor alone can meet people's currentneeds. The intervention of a new type of space may completely break the authenticity here. But from another perspective, if authenticity must be very important? is the narrative logic behind it of other intentions?

吉屋出租
FOR RENT

邻近师爷馆的「出租」告示。自从徐渭艺术馆正式对公众开放之后，更多老屋被改造成咖啡店、茶室和小卖铺。有关商业空间的想象，或许还有更多可能性。

青藤片区｜2021-09-08

旅途愉快
ENJOY YOUR TRIP

大乘弄的青藤书屋正对面,
木凳、晾衣架,再加一张白纸黑字的说明,完美。
摇着蒲扇的叔叔隐身在黑暗之中。

青藤书屋对面 | 2021-09-08

这家仓桥直街的咖啡厅已经焕新,但走过的路人大多还是生活在这儿的旧人。老人更喜欢在街上肆无忌惮没有口罩阻隔地聊天。年轻人戴着口罩进了咖啡馆。不同时代、不同年龄人的生活规律,会逐渐在这条街上同时呈现。

生活一直在不断改变。

Although cafés in Cangqiaozhi Street have renewed to cater for young fashions, most of the passers-by are still the aged residents who prefer to meet and chat in the street without the barrier of masks. Young people entering the café are accustomed to the rules with masks on. Gradually the routines of different generations and ages happen on this street at the same time.

The life here is continually evolving.

榴花斋
LIU HUA ZHAI

榴花斋的更新重建正在进行中。

九月的徐渭艺术馆进入闭馆期，门口的广场非常安静。

青藤广场 | 2021-09-08

优雅
ELEGANCE

院子中央坐落着几处老民居,形成独特的居住空间。

与窗子里的猫咪完成社交的"黄尾巴",此时正在散步。

后观巷附近居民楼 | 2021-09-08

你瞅啥
WHAT'S UR PROBLEM?

后观巷东边入口处的一栋六层居民楼里,
一只猫咪正在社交。

窗子向北,面朝院子。

青后观巷附近居民楼 | 2021-09-08

第九章
CHAPTER
09

看！"灰机"
美发
窗框
地（上的）瓜
熙熙攘攘
其实我
恰如其分

LET ME HAVE A LOOK LOOK
HAIRDRESSING
REEMPLOYMENT
SWEET POTATOES
TEN SECONDS BUSTLING
'I LOOK LIKE A PIG'
JUST RIGHT
DAILY GREETING

徐渭艺术馆老墙 | 2021-10-27

第九章　CHAPTER 09

还记得"双十一"诞生之初,是单身用来自嘲、盼着剑走偏锋成功"脱单"的吗?

从这一天正式被阿里巴巴招募的 2009 年,至今十几年过去。改头换面后的"光棍节"闪亮登场,令所有人大开眼界,人们欢天喜地迈入消费主义重金打造的"梦幻嘉年华"。

而十几年后的今天,"光棍儿"的数量显然没少,单身也正在破除污名化,生育率也是持续走低。不过,今宵有酒今宵醉。我们依然需要面对那个老问题:如何生活?

仓桥直街上虚掩的门缝里闪着霓虹色的光,女主播卖力工作的声音隐约可闻;萝卜、红薯纷纷出动,食物再次占领街道;也许真正释怀的只有那些毛茸茸的小家伙,顶着"单身汪"的污名活得倒是很有节奏。

Do you know that the day of 11 November was set down first by singles for self-mockery in the hope of successfully being taken by his or her most significant other?

Over ten years have passed since the day was officially used as 'Double 11 Shopping Spree' by Alibaba in 2009. The shopping extravaganza falling on that day was unprecedent, dazzling those singles, and fascinating all people to walk into 'the consumer carnival in delight'.

The past over ten years has witnessed the steady number of bachelors and bachelorettes, the broken-down stigma of being single, and the declining fertility rate. As young people now believe love may come and love may go, they still have to face the same meta-question: how to live?

The neon lighting glows from the hidden doors of houses in Cangqiaozhi Street with the voice of the camgirl proving her utmost at work. Food is once again taking over the street as cooked turnips and sweet potatoes are popular here. Perhaps the only ones who are really relieved are the fluffy little ones who unexpectedly inflicted on stigma of naming single persons 'single dog'.

看！"灰机"
LET ME HAVE A LOOK LOOK

一年一度最长假期的结束，还给青藤片区一段安静时光。

为了记录师爷馆现状，"小飞机"正在空中作业。

绍兴师爷馆门口 | 2021-10-27

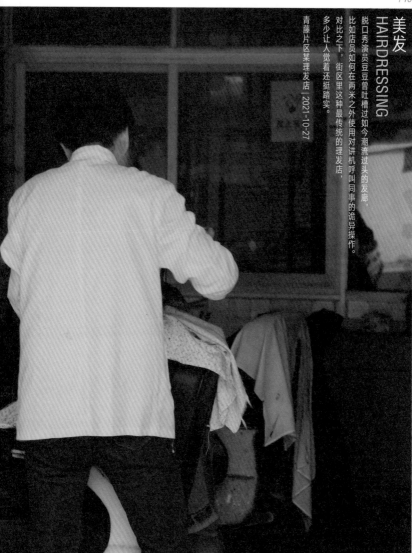

美发
HAIRDRESSING

脱口秀演员豆豆曾吐槽过如今潮流过头的发廊，比如店员如何在两米之外使用对讲机呼叫同事的诡异操作。对比之下，街区里这种最传统的理发店，多少让人觉着还挺踏实。

青藤片区某理发店 | 2021-10-27

窗框再就业
REEMPLOYMENT

食物成为一种特殊的介质,朴素地连结着人与时空。
在当下这个微小的场域中,窗框与萝卜正在上演神奇魔法幻术。

青藤片区 | 2021-10-27

食物是季节属性最明显的东西，腊肉、萝卜干、小鱼干、红薯……绍兴这个四季不是很分明的江南小城，食物成为时间与空间最紧密的纽带，地瓜连接起了仓桥直街与绍兴的秋天。法国哲学家吉尔·德勒斯提出过"块茎思维"，为建筑师在解决空间形态问题上提供了哲学支持，仓桥直街每一处季节食物的变化也通过一个个点状的"块茎"，建立起强连接，这背后每一个"块茎"都是一个点，没有开始，没有结束。它逐渐生长，日益蓬勃，青藤不再是青藤，是又见青藤。

Food is of the most obvious seasonal attributes. Cured meat, dried radish, dried small fish, sweet potato... Shaoxing, a small town in the south of the Yangtze River without distinct four seasons, nurtures such food, which has become the closest link between time and space. Sweet potato bridges Cangqiaozhi Street and Autumn in Shaoxing. Just like the rhizome metaphor proposed French philosopher Gilles Deleuze, the architect provided philosophical support in dealing with the spatial form, and the changes of seasonal food in Cangqiaozhi Street stem from countless pointshaped "rhizome" via strong links. Behind each "rhizome" is a point, without beginning and end. The community grows gradually and becomes more vigorous. It is no longer Qingteng, but 'Seeing Qingteng Again'.

地（上的）瓜
SWEET POTATOES

红薯可谓是"南北通吃型"食物，比如：凛冽中看它们从黑黢黢的大罐子里被取出来时，空气里弥散着的幸福泡泡；或者，周末午后的嘴巴里嚼着红薯干，烤着电暖气抵御南方冬日的阴冷。

仓桥直街 | 2021-10-27

十秒熙攘
TEN SECONDS BUSTLING

混乱只发生了十秒钟,四人三狗。

狂吠的狗不满于拍照的入侵,饥渴的相机飞快按着快门,主人慌乱之中只能露出些许尴尬的腼腆微笑。

青藤片区 | 2021-10-27

"其实我是猪"
'I LOOK LIKE A PIG'

门口的白绒绒,似乎躺出了一种生活态度。

拍照时,也不忘记友好地摇着尾巴互动,动作悠然,看来是绍兴古城的速度。

后观巷 | 2021-10-27

旧码头对面的老房子经过立面的粉刷，使巷子看起来干净不少。吃完早点，老奶奶倚门而立，隔壁邻居消食路过聊了起来。竟让我想到古龙的一本小说 再劣的茶，只要热喝就能下口——就像女孩子只要年轻，就总是可爱的。后半句肯定是不对的，但热茶、冰啤酒、冰可乐一定是好喝的！所以在老街区与旧房子中，只要人与人之间很近就是生活，就是温暖幸福的生活。很多住在青藤的老爷爷、老奶奶都不愿离开，因为这里街道很窄，离生活很近。

The newly painted facades of old houses across from the old pier refresh the alleyway with a much neater appearance. After breakfast, the old lady leans against the door while the next-door neighbour dropped by chatted after meal. It reminds me of a saying by the novelist Gu Long that even the cheapest tea may lure people with fragrance as long as it is hot, it is the same with girls whose lovely souls will not be wrinkled by time. The latter half of the sentence seems rude, but it cannot be denied that you cannot never miss the delicious moment tasting the hot tea, cold beer and cold coke! In the old neighbourhoods of old houses, life is the closeness between people. Such liking and loving over the long term make the life in Qingteng Community warm and happy. That is why many elders living here never think about move, for the streets are shared space for life.

日常问候
DAILY GREETING

奶奶身子倚着门框,微微笑瞅着街道,不多时便瞅到了熟人。

至于她俩在聊什么,我们认真地瞎猜了一番。

仓桥直街 | 2021-10-27

项目介绍 PROJECTS IN QINGTENG SHAOXING

1 徐渭艺术馆及青藤广场
Xu Wei Art Museum and Qingteng Square

2 青藤别苑
QINGTEGN BIEYUAN

3 张家台门 ZHANG JIA TAIMEN

5 绍兴师爷馆 SHAOXING SHIYE EXHIBITION

4 榴花斋 liu hua zhai

设计团队 DESIGN GROUPS

主创建筑师
胡慧峰

设计团队
蒋兰兰　章晨帆　韩立帆　朱金运　李鹏飞　黄迪奇

委托方
绍兴市文化旅游集团

施工方
浙江勤业建工集团有限公司

合作方
浙江星睿幕墙装饰工程有限公司版
故宫出版社
广东集美设计工程有限公司联合体

青藤手记

结构专业
张 杰 陈 旭 吕君锋 丁子文 沈泽平 陈晓东

水专业
易家松 邵煜然

暖通专业
潘大红 李咏梅

电气专业
张 薇 俞 良 杜枝枝

弱电专业
林 华 叶敏捷 杨国忠

景观专业
吴维凌 王洁涛 吴 敌 朱 靖 敖丹丹 何 颖 林 腾

室内专业
楚 冉 刘婉琳 汪军政 梅文斌

展陈专业
赵同庆 梁 爽 陈 伟 黄世琰 孙小童

照明专业
王小冬 赵艳秋 傅东明 冯百乐 吴旭辉

幕墙专业
史炯炯 王皆能 段羽壮 张 杰

基坑围护专业
徐铨彪 曾 凯

BIM 设计
张顺进 任 伟 严宜涛 王启波

EPC
房朝君 王 青 苗 赛 贝思伽 李延琦 李 晨

主创建筑师

蒋兰兰

浙江大学建筑设计研究院
建筑创作研究中心副主任
建筑学硕士
国家一级注册建筑师
高级工程师
杭州市优秀青年建筑师

胡慧峰

浙江大学建筑设计研究院总建筑师
建筑创作研究中心主任
研究员，国家一级注册建筑师
中国建筑学会青年建筑师奖获得者
浙江省工程勘察设计大师

青藤手记

章晨帆

浙江大学建筑设计研究院
建筑创作研究中心主任助理
建筑学硕士
高级工程师

立帆

⋯⋯大学建筑设计研究院
⋯筑学硕士
⋯级工程师

朱金运

浙江大学建筑设计研究院
建筑学学士
中级工程师

李鹏飞

浙江大学建筑设计研究院
建筑学硕士
中级工程师

黄迪奇

浙江大学建筑设计研究院
建筑学硕士
中级工程师

后记 EPILOGUE

春夏秋冬,周而复始,青藤在我们每个人眼中也带着不同的色彩。有宣纸泼墨的黑白、夜里灯光泛着的淡黄、春夏青藤盛开的紫、乌篷码头洗衣荡起的青、腊肉晒干的棕红。

日日驻场的建筑师已经转向了热火朝天的阳明工地,书圣故里和八字桥都能看到我们的身影。建筑工人和推土机撤离青藤,安静整洁的街区咖啡厅、茶室、民宿逐渐开了张,青藤书屋的游客数量大概超过了过去五年的总和,朝气蓬勃的面孔越来越多,广场越来越热闹,有人离开,也有人到来。

我们发起了"又见青藤",但我们"不止青藤"。

Spring, summer, autumn and winter- the circles of life turn, and we turn with a whole year recording of Notes on Qingteng, as the community is about to celebrate another new year. This time, we have many stories. Since there are a thousand Hamlets in a thousand people's eyes, each of our eyes perceive those quotidian existence in different colours, such as the black and white of ink on Chinese rice paper, the luminous pale yellow at night, the blossoming purple in spring and summer, the rippling green when doing laundry at Wupeng Boat pier, and the brownish red of the cured meat.

Yangming site, including Home of the Sage of Calligraphy and the Baziqiao Bridges, is in full swing and crowed with our figures, as site architects carried daily commissions there while construction workers and bulldozers withdrew from Qingteng. The quiet and tidy community once again embraces cafes, tea houses and B&Bs, as well as tourists to Qingteng Study whose number probably exceeds the visitor totals for the past five years. Among them, increasing faces are young and vibrant. The rushing crowd also makes the square more bustling.

We launched 'YOU JIAN QINGTENG', but it is with knobs on.

by UAD ACRC